Brunel
Royal Albe₁
Saltash – Revisited

Thomas Bowden
ATM Publishing

An engraving of the Bridge as viewed from Saltash shortly after opening in 1859.   ***H. Adlard.***

This book is dedicated to my wife, Angela, without whose help and encouragement it would not have been written.

Thomas Bowden

Index

Brunel the Engineer	1
Crossing the Tamar	4
Construction – The Centre Pier	6
Construction – The Bridge	18
Floating & Raising the Trusses	25
The Opening Ceremony	33
The Bridge 1859 – 2009	36
Signalling the Bridge	51
The Bridge Builders	54
Tales of the Bridge	60
Viewing the Bridge	62
Epilogue	65
Bibliography	68
Acknowledgments	72

Introduction

I have had a lifelong interest in the Bridge since I was four years old and now over fifty years later I still experience a sense of excitement whenever I see the Bridge, especially when I consider how it was designed and built. This has prompted me to write this volume to commemorate the 150th anniversary of the opening of the Bridge by His Royal Highness Prince Albert the Prince Consort on 2nd. May 1859. It incorporates additional information which has come to light since I first wrote about the Bridge in 1983. My first book was called "Brunel's Royal Albert Bridge, Saltash" hence the title of this book, "Brunel's Royal Albert Bridge Saltash – Revisited". The first book was co-authored with my old friend Bernard Mills. However, since then Bernard has become a railway author in his own right, his only involvement with this volume has been the chapter on "Signalling the Bridge".

I have incorporated into this book a new section on the builders of the Bridge in order to give some background information on the personalities and Contractors involved with the construction of this magnificent structure.

Please note, all measurements shown are in Imperial Units as these were the ones in use when the Bridge was built.

Thomas Bowden. Ivybridge. Devon. April 2009.

The Seal of the Cornwall Railway Company.
W.J. Power collection.

Brunel the Engineer

Marc Isambard Brunel was a French Royalist who had fled from the French Revolution, firstly to America where he became Chief Engineer of New York, and then to England where he married Sophia Kingdom daughter of William Kingdom, a Plymouth Naval Contractor in 1799.

Marc designed and installed the world's first ships' block making machinery for the Royal Navy in Portsmouth Dockyard. Under these circumstances it was necessary for the Brunels to reside as near as possible to the Dockyard in order to supervise the installation of the new machinery. They moved to a house in Portsea, Portsmouth where Isambard Kingdom Brunel was born on 9th April 1806. After the machinery had been installed, the Brunels moved to London where they lived in a house in Lindsay Row, Chelsea before moving to Bridge Street, Blackfriars when Marc started work on the Thames Tunnel.

It was whilst the Brunels were living in Chelsea that Isambard received his first education from the Reverend Weedon Butler who lived nearby. Isambard also received tuition from his father and by the time he was six had become proficient in mathematics, geometry and drawing. He was then sent to a preparatory school in Hove where he amused himself in his spare time by drawing a map of the town. When completed, he went on to draw some of the more prominent buildings. Indeed, in one of his letters home he asked his father to send him his 80 foot long measure so that he could prepare some scale drawings of these buildings. Whilst at Hove he amazed some of his school friends by predicting the partial collapse of some buildings that were being erected across the road from the school. Overnight there was a storm and the buildings collapsed as Isambard had predicted.

In order to complete his education, Isambard was sent to France in November 1820, firstly to the College of Henri Quatre, Paris for a period of nearly two years. Then, his father arranged for him to become apprenticed to Louis Breguet who was a famous maker of chronometers, watches and scientific instruments. On his return from France Isambard found employment in the office of his father who was engaged in the construction of the Thames Tunnel.

At the age of 24, Isambard won a competition to design a bridge across the Avon Gorge at Clifton, Bristol but as fate decreed, his Clifton Suspension Bridge was not completed, due to financial difficulties, until 1864 five years after his death.

In 1833, aged 27, he was appointed Engineer to the Great Western Railway, then being planned to connect Bristol with London. The steepest part of this line is east of Bath where there is a falling gradient towards Bath through the tunnel at Box. The tunnel is nearly 2 miles long and was built on a falling gradient of 1 in 100. Work had started on the tunnel by September 1836 and it was not completed until June 1841. For over two and a half years one ton of candles and one ton of gunpowder were used every week. Over thirty million bricks were used which were all manufactured locally. Unfortunately, 100 men lost their lives during the construction of the tunnel, compare this to the construction of the Royal Albert Bridge where not a single life was lost.

During his short life, for he died on 15th. September 1859, Brunel had been responsible for the construction of many railways, three ocean going steamships, a portable Military Hospital for the Crimean War, the Clifton Suspension Bridge at Bristol, as well as the Atmospheric Railway in South Devon plus many more minor projects. He was also on the Committee of the Great Exhibition which opened in London and Vice President of the Instituition of Civil Engineers.

Many of Brunel's works are still in existence today, 150 years after his death, a testament to his skills as an engineer. Brunel's last masterpiece, the Royal Albert Bridge at Saltash is the subject of this book.

Isambard Kingdom Brunel from a painting by J. Horsley R.A.
T.N. Bowden collection.

Crossing the Tamar

In August 1843 the promoters of the Cornwall Railway asked Captain William Scarfe Moorsom to survey a possible route for a railway from Falmouth to Plymouth. Moorsom's survey was completed by September 1844. He proposed to cross the River Tamar by means of a steam train ferry at Torpoint where the river was just over a half mile wide. This plan was submitted to Parliament for their approval and was thrown out by the House of Lords in May 1845.

The promoters of the Cornwall Railway then engaged Brunel as their Engineer to succeed Moorsom. Brunel's first plan was for a steam train ferry at Saltash. However, under it Act, the Cornwall Railway was only allowed to cross the Tamar by means of a bridge. Following the rejection of the ferry plan Brunel's next idea was for a timber bridge of seven spans, being one of 255 feet and six of 105 feet, giving a clear headway above high water of 80 feet, this was rejected by the Admiralty.

The Cornwall Railway obtained its Act, in August 1846. Two of the stipulations were that the river could only be crossed by means of a bridge of not more than four spans and approval of any design and the materials used for its construction had to be obtained from the Lord High Admiral.

Brunel then submitted a modified plan of two spans of 300 feet and two of 200 feet with a clear headway above high water of 100 feet. Once again the Admiralty rejected the plans and went on to stipulate that there should not be more than one pier in the fairway and that a minimum height of 100 feet above high water was required. Also, the bridge was to have flat soffits.

In view of these stipulations Brunel abandoned all plans for a double track timber structure and instead concentrated on a single track wrought iron bridge with one span of 850 feet, which would have weighed 7,000 tons and cost £250,000 to erect at 1846 prices. Enquiries were made as to whether it was possible to construct such a huge span, and what means were available to lift the span to its final position 100 feet above the Tamar. The resulting answers indicated that it was feasible but the costs were considered to be prohibitive.

Brunel then went on to prepare a second design for a wrought iron bridge which had two spans of 465 feet (later reduced to 455 feet) along with seventeen land spans varying in length between 70 and 90 feet. Only one pier would be required to stand in the fairway and a clearance of 100 feet above the water would be provided. Both the Admiralty and the Directors of the Cornwall Railway gave their approval to this plan. However, the late 1840's were very difficult times in which to raise capital for such projects, so Brunel then approached the Board of Trade for permission to alter the design to a single track bridge as this would save £100,000. Approval was granted, work could now commence.

Brunel had largely finalised the design for the Bridge by October 1852. The contract for the construction of the bridge was let to C.J. Mare of Blackwall, London for the sum of £162,000, which at 2009 prices is roughly £16.64 million. On 1st. October 1853 C.J. Mare signed the contract. Interestingly, under the terms of the contract there was a clause which said that, "If the contractor should go bankrupt during the construction of the Bridge, the works would revert to the Cornwall Railway Company and Brunel, for them to complete by means of direct labour or by engaging another contractor". We shall see the importance of this clause later in the book.

As one of the stipulations contained within its Act, the Cornwall Railway Company or its successors have to pay the Duchy of Cornwall an annual rent of £25 per annum in perpetuity (1846 prices). At 2009 prices this is the equivalent of £1,734!

Construction –
The Centre Pier

Once his plans had been approved in 1848 Brunel set about planning the exploration of the bed of the River Tamar, with a view of founding a deep water pier midstream. The rock, a very hard form of Greenstone Trap, lay beneath 80 feet of water, mud and dense oyster beds, Saltash at one time having enjoyed a thriving oyster industry.

He decided that the only way was to construct a huge wrought iron cylinder 85 feet long and 6 feet in diameter, which was duly built and floated out in two halves, bolted together midstream and then lowered vertically until it reached the river bed. The mud sealed the lower edge of the cylinder which allowed the residual mud and water to be pumped out. This permitted trial borings to be made into the rock below in order to determine whether it was suitable for the founding of a deep water pier. Altogether the cylinder was raised and lowered a total of 35 times and a total of 175 borings made, enabling Brunel to construct an accurate model of the invisible rock below. As a result of these tests, it was clear that the rock was capable of supporting a pier.

In order to prove the feasibility of this scheme, Brunel had a piece of trial masonry built on the rock. Embedded in the masonry was a wooden box containing a copper plate bearing the inscription, *'Cornwall Railway, Saltash Bridge, Trial Foundation of Central Pier, January 1849. I.K. Brunel, Engineer, William Glennie, Resident Engineer'.* However, through a lack of funds further work was suspended for a period of three years.

Brunel's Assistant, Robert Pearson Brereton delivered a paper in 1862 to the Institution of Civil Engineers which gave a detailed account of the construction of the centre pier. A copy of this paper follows:

March 4, 1862.
JOHN HAWKSHAW, President,
in the Chair.

No. 1,062._ "Description of the Centre Pier of the Saltash Bridge on the Cornwall Railway, and the means employed for its Construction." by ROBERT PEARSON BRERETON, M. Inst. C.E.

The Author having been frequently requested to furnish a description of the means employed for the construction of the centre, or deep-water pier of the Albert Bridge, across the River Tamar at Saltash, on the design and construction of which he was engaged under the late Mr. Brunel (V.P. Inst. C.E.), the Engineer of the Cornwall Railway, has prepared the following narrative:-

"In the year 1845, the Cornwall Railway Company applied to Parliament for an Act to construct a Railway from Plymouth to Falmouth. The locality selected for crossing the River Tamar was at the town of Saltash, about 3 miles north of Plymouth, at a place where the river narrows to 1,100 feet wide, with precipitious banks, and has a depth of 70 feet, from the surface of the water to that of the mud.

It was at first proposed, that the bridge should consist of seven openings, one of 250 feet, and six of 100 feet each, of a uniform height of 70 feet; but in consequence of objections raised by the parties interested in the navigation, plans were submitted to the Admiralty with the increased dimensions of one span of 255 feet, and six of 105 feet, at a height of 80 feet above high water, all to be built with timber-trussed arches. The Admiralty, however, required that there should be four spans only, two of 300 feet and two of 200 feet each, with straight soffits, and a clear headway above high water of 100 feet. To comply with these requirements, it became necessary to apply to Parliament, for amended powers, which were obtained in 1847.

In selecting the site for the bridge, advantage was taken of a dyke of greenstone trap, which intersects the clay slate formation in the neighbourhood, and crops out to the surface above water on the

7

Saltash, or western bank of the river. In 1847, a general examination of the bed of the river was made, by borings. It was thus ascertained, that along the eastern side, to beyond the middle of the channel, there was rock, covered with a deposit of mud, or silt fom 3 feet to 16 feet in thickness; and that from the middle of the channel the rock fell off rapidly towards the Saltash side. Attempts were made to ascertain, by borings, the figure of the surface of the rock below the silt, in the middle of the river; but owing to the depth of the water and the strength of the current, sufficiently reliable information could not be obtained. It was discovered by divers, that on the Saltash side, the surface of the rock presented a precipitious decline towards the river, but that it was free from chasms, and in the inshore part was favourable for the construction of a pier. In 1848, Mr Brunel determined on a thorough examination of that part of the bed of the river, where the pier would probably be built. This was effected by means of a cylinder of wrought iron, 6 feet diameter, and 85 feet long, which was slung between two gun-brig hulks moored in the river, and the bottom edge of which was lowered a few feet into the mud. Early in 1849, the bed of the river for a space of 50 feet square, had been carefully examined by one hundred and seventy-five borings, made inside the cylinder at the thirty-five places where it had been pitched. On the last occasion, the cylinder was sunk to the rock, the water was pumped out, to test the water-tightness of the mud, and the material was excavated from the inside. Masonry was then built upon the rock, up to the level of the bed of the river, and the cylinder was withdrawn. From the information thus obtained, an exact model of the surface of the rock, which had been examined, was prepared, showing the irregularities and the fissures that were to be expected.

By this time the construction of wrought-iron railway bridges had become general; and in the cases of the Conway and the Britannia Bridges, spans of up to 460 feet had been obtained without difficulty. In determining the dimensions of the bridge at Saltash, it was considered whether it might not, with advantage, be constructed with only one pier in the deep water, instead of three, as would have been necessary for the spans required by the

Admiralty; and the experience obtained of the nature of the foundations, having shown that the rock was favourable for the construction of the piers fo a span of 465 feet over the western half of the river, designs were made in 1850 for a bridge with two main spans of 465 feet each. In 1852, when it was determined to proceed with the building of the bridge, it was considered practicable to reduce the spans to 455 feet each. The drawings from which the bridge was executed were prepared accordingly, and early in 1853 the work was commenced.

The total length of the bridge, including the adjoining land openings, is 2,200 feet. It has two spans of 455 feet, two of 93 feet, two of 83 feet 6 inches, two of 78 feet, two of 72 feet 6 inches, and nine of 69 feet 6 inches. The centre, or deep water pier, which carries the weight of one half of each of the two main spans, consists of a column, or circular pillar, of solid masonry, 35 feet diameter, and 96 feet high from the rock formation to above high-water mark. Upon this are placed four octagonal columns of cast iron, 10 feet diameter, carried up to the level of the roadway, which is 100 feet above high-water mark. Upon the tops of these columns, cast-iron standards are fixed, for receiving the ends of the tubes and chains, which constitute the trusses of the bridge. The weight at the bottom of the masonry foundation is about 9 ½ tons per square foot, increased, when the bridge is loaded by passing trains, to about 10 tons per square foot.

For the construction of the masonry pier, a wrought-iron cylinder, of boiler plates, 37 feet diameter and 90 feet in length, and open at the bottom, was sunk through the mud of the bed of the river to the rock. The water was afterwards pumped out, and the mud excavated; the masonry columns being built up inside, and the cylinder above the ground being afterwards removed. From the experience obtained in sinking the 6-feet cylinder for making the borings, it was expected that, with the compression of the mud outside the 37-feet cylinder, by a bank thrown round it, after it was sunk to the rock, it would be rendered sufficiently water-tight for the execution of the masonry. But to provide for the contingency of excessive leakage, it was determined, in designing the cylinder, so to construct it as to admit of the application of air

9

pressure in the inside, which had been successfully employed, in sinking the cylinders, for the bridges at Rochester and at Chepstow. The shape of the bottom of the cylinder was obtained, by applying to it a model of the surface of the rock, which had been prepared from the previous borings. This surface, although very irregular and ragged, showed a general dip, of about 6 feet, to the south-west, and the bottom of the cylinder was formed with a corresponding slope, one side being 6 feet longer than the other. A dome, or lower deck, was constructed inside, at the level at which the mud of the river bed would be, when the cylinder was sunk in position; and an internal cylinder, 10 feet in diameter, open at the top and the bottom, connected the dome with the top, or upper deck, of the cylinder. Further to allow the application of air pressure, the 6-feet cylinder, previously used for boring, was fixed excentrically inside the 10-feet cylinder, and an air-jacket, or gallery, making an inner skin round the bottom edge below the dome, was formed of plates, intended to be used in the roadway girders of the bridge. The air-jacket was about 4 feet in width, divided by partition plates into eleven compartments, or cells, and connected with the bottom of the 6-feet cylinder by an air passage below the dome. It was intended, that the lower part of the cylinder, below the ground, should be filled in with solid masonry, and that it should not be withdrawn. The upper part, 50 feet in length, was constructed in two pieces of larger diameter than the masonry, with vertical separating joints bolted together, so as to be capable of being removed after completion of the pier.

In the spring of 1853, the cylinder was commenced. It was built upon the beach, or shore above the river, with its lower end towards the water, and with its axis at an inclination of about 1 in 6. It was designed to float at that inclination when empty, and to draw about 15 feet of water forward. Whilst the cylinder was being constructed, four wrought-iron pontoons, intended for floating the large spans of the bridge into their places, were moored in the middle of the river, round the intended site of the centre pier, and four mooring anchors and cables were laid out, ready to receive the cylinder.

In May 1854, the cylinder was launched, by being hauled down to low-water mark on the launching ways, and it was floated off by the rising tide. It was then towed to the pontoons, and placed between them, and some water was admitted to bring it into its proper position, with the upper deck well out of the water. The mooring chains were taken on board, rove through sheaves, fixed at four points low down in the cylinder, and brought to the purchases on the deck. After the cylinder was secured between the four pontoons, water was gradually let in, till it floated in a vertical position. It was then brought to the intended site, and sunk early in June, by gradually admitting water to the inside. The cylinder penetrated about 13 feet through the mud at the bottom of the river, and landing on some irregularities upon the rock, its top heeled over from the vertical, about 7 feet 6 inches towards the east. Preparations were then made for resorting to the use of air pressure, in order to reach the bottom edge, and to bring the cylinder into an upright position. A pneumatic apparatus and air pumps, which had been used at the Chepstow Bridge, were obtained, and fixed to the top of the 6-feet cylinder; two engines of 10 H.P. were placed upon the upper deck to work the air pumps, and two 13-inch water pumps were fixed inside the 10-feet cylinder. Meanwhile, steps were taken for forcing down and righting the cylinder, by water pressure upon the dome, or lower deck, and by loading the higher side with iron ballast. Gravel was thrown round the cylinder, to secure the surface of the mud from scour, but the quantity was too small to interfere with the "righting" of the cylinder.

At the end of June, the cylinder forced its way through the obstructions at the bottom edge, and went down 3 feet, taking a nearly upright position. Early in July, the air and water pumps were set to work. In August, the greater part of the mud and oyster shells, which filled the compartments of the air-jacket at the bottom, had been cleared out, and the irregular surface of the rock was being excavated. The water pumps were used for lowering the surface of the water inside the 10-feet cylinder, thereby diminishing the pressure in which the men were obliged to work; the bottom of the cylinder being then 82 feet below the high-water level.

Between August and November, a leak having broken out, through a fissure in the rock on the north-east, or higher edge, great difficulty was experienced, in maintaining sufficient pressure with the air-pumps, to keep the water down, and the bottom dry, for the men to work, while excavating the rock. The leakage was at length considerably reduced, by driving close sheet piling into the fissure.

In February 1855, the cylinder was sunk to its full depth, in an upright position; its bottom being everywhere down to the rock, and 87 feet 6 inches below high water at its lowest place. Before the final dressing of the rock at the bottom, a hemp gasket was worked under the edge of the cylinder, all round the outside, to assist in making it water-tight. The bottom edge, both on the up and on the down stream sides, was also secured to the rock, by Lewis bolts, so as to steady it against the action of the current and the tides. In March, the rock was levelled in all the compartments of the air-jacket, and its surface cemented over. The cutting away of the rock was a very tedious operation, as in some parts, as much as 6 feet in depth had to be taken off with chisels, the trap being of such hardness, that tools could with difficulty be got to work on it. In April, a ring of granite ashlar masonry was commenced in the air-jacket, and in May, it was completed all round. The granite ring was 4 feet thick, and averaged about 7 feet in height. Whilst the masonry was being proceeded with, a bank, about 10 feet high, of heavy material, composed of slag and sand, was thrown round the outside of the cylinder, to compress the mud. The operations with the air-jacket were thus completed, the greatest pressure having been with 86 feet head of water. From thirty to forty men were engaged inside.

Early in June, the air-apparatus was removed from the 6-feet cylinder, the water was pumped out of the body of the cylinder below the dome, and the excavation of the mud had been commenced; but at the end of the month a leak broke out. As the pumps were unable to remove the water, two additional engines and 13-inch pumps were provided, and in the interval, until September, efforts were made to diminish the leakage, with varying success, by throwing more material round the bank outside the cylinder. The leakage, however, continued to such an extent; that even with the two additional pumps, which were got to work in October, the water could not be kept down. In November, it still required the four pumps to keep the water down to 54 feet, inspite of incessant pumping and throwing additional material round the outside. It was therefore thought, that recourse be had to air pressure in the body of the cylinder below the dome, and preparations for its application were made.

As the 10-feet cylinder, for 50 feet in length at its upper part, was not constructed with plates of sufficient thickness to support, with safety, the pressure when filled with air, a 9-feet cylinder, with thicker plates, and about 50 feet in length, was constructed to be slipped inside, and to be secured at the bottom to the 10-feet cylinder, by a shelf riveted to it, about 50 feet from the top, to which level the four pumps were able to keep the water. Two pairs of new 12-inch air-pumps were provided, in addition to those hitherto used, and commodious wrought-iron air locks, or cages, were constructed, for attaching to the top of the 9-feet cylinder. By the middle of November the shelf was fixed, and the 6-feet cylinder was removed.

To provide against the buoyancy, or upward pressure against the dome and cover of the 9-feet cylinder, it was necessary to load the 37-feet cylinder, in addition to its own weight of 290 tons, with about 750 tons of ballast. Of this about 350 tons consisted of pig iron and kentledge, stacked upon the shelves of the 37-feet cylinder, near the top, and upon cross girders under the upper deck. The remaining 400 tons were placed upon the dome; 100 tons of this consisted of sand in bags lowered through the water, and uniformly distributed, and 300 tons of pig iron, afterwards thrown in upon it. These operations were not completed until the middle of December, when everything was ready for fixing the 9-feet cylinder. The pumps were then brought into good order, and by continued pumping, the water was kept down. The mud was excavated, the cylinder below the dome securely shored across, and the rock levelled. The masonry in the body of the cylinder was commenced early in January 1856. A leak was discovered to have again broken out at the same fissure as before, and it had torn away the rock underneath the masonry of the airjacket and the bottom edge of the cylinder; but the masonry itself was not disturbed.

As soon as the rock was reached, holding-down Lewis bolts were let into it, with iron bars to be built into the masonry, and attached to the bottom of the 10-feet cylinder, as a further precaution in case of any sudden influx of water. The water from the leak, and from some smaller fissures, was carefully collected in pipes, and conveyed into two cast-iron wells, which were formed in halves and placed round the suctions of the water-pumps, the whole being then built into the solid masonry.

By incessant pumping, the water was kept down, so that by the end of February, the masonry, in thin courses of granite ashlar in cement, had reached the level of the air-jacket ring. The masonry was then thoroughly bonded together, the plates of the air-jacket being cut out as it proceeded. Upon the top of the bonding-course, two courses of hard brickwork in cement were laid, making a perfectly water-tight floor over the whole surface of the section, which was there 35 feet in diameter. Meanwhile, the masonry of the air-jacket, where the leak occurred, was taken down, and the leak was diminished by additional sheet piling at the edge of the cylinder, so that one engine and pump could be dispensed with. The water was collected in a pipe having a sluice valve, which on being closed early in March reduced the leakage by about one third. In the middle of March the water was drawn off above the dome, the ballast was removed, and the 37-feet cylinder above it properly shored. Early in April, the masonry had reached the springing level of the dome, or 20 feet in height. The dome was then cut away, as well as the bottom of the 10-feet cylinder. Two pumps were now sufficient to keep the water down.

By the middle of April, one of the pump-wells was filled up with cement concrete. Early in June, one pump was found sufficient to keep the water down, the masonry being 43 feet high, and proceeding at from 5 feet to 7 feet in height per week. In the middle of June, the masonry being then 46 feet in height, the influx of water was entirely stopped, by filling the remaining pump-well with cement concrete, and then closing the top of it.

14

The masonry was completed to the top of the plinth, or cap of the pier, by the end of October; and in the middle of November, the upper part of the cylinder was unbolted, at the separating joints, and floated to the shore with the pontoons, and the iron was made use of, for the decking of the pontoons which were to be used in floating the bridge

All that then remained to be done, was to fix the bases of the cast-iron columns to the masonry, and to erect the columns themselves for receiving the ends of the bridge when floated into its place. This was accomplished, with the first span, over the western half of the river, in September, 1857; and with the second, or eastern half, in July, 1858."

At first, seven hour shifts were worked at the bottom of the cylinder, however, some men were suffering from partial paralysis and it was taking them two or three days to recover from what is now known as the "bends". It was found that the men suffered no such problems if they worked three hour shifts at the bottom of the cylinder. It was also found that Englishmen were more able to stand the pressure than Irishmen due to their more robust frame and better standard of living.

The Great Cylinder being floated out in May 1854. *T.N. Bowden collection.*

15

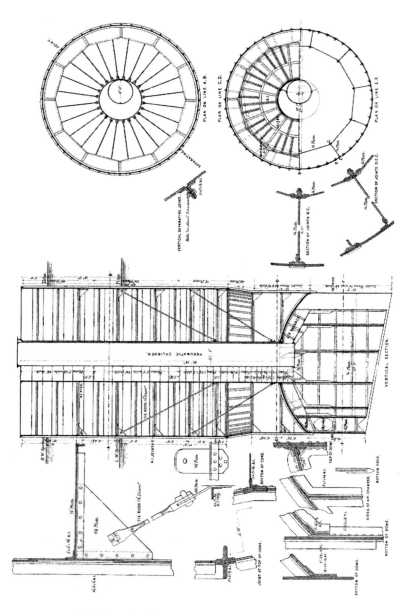

These two drawings show the arrangement of the Great Cylinder used for the construction of the centre pier. *R.P. Brereton.*

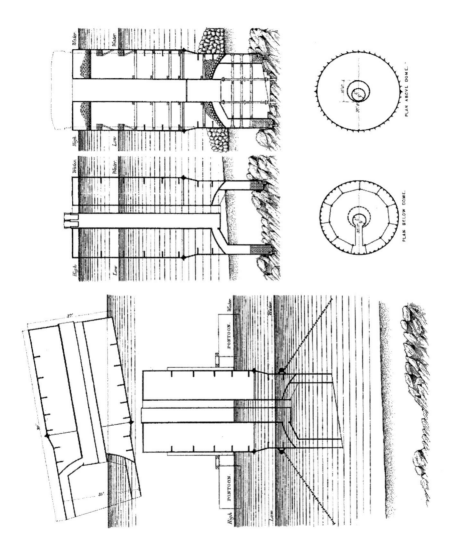

Construction – The Bridge

Whilst work was progressing on the centre pier, the contractor C.J. Mare of Blackwall was establishing his workshops on the Devon shore of the river. The works were known as "Saltash Bridge Works" and were situated north of what is now the "Royal Albert Bridge" Inn. Several buildings were demolished to make way for the Works however, the Inn was not one of them! A jetty was built to allow materials to be brought in by water. Land was cleared under what is now the Royal Albert and Tamar Bridges to allow the two great trusses to be built. Deep piling was installed on the foreshore to take the weight of the completed trusses which had an estimated weight of 1,060 tons each. Whilst it was being built, the tube was supported on timber scaffolding which completely enveloped the truss. From the Saltash side of the river it must have looked as though the truss was being built of wood not iron!

Each truss is neither a true arch nor a true suspension structure. The Bridge is believed to be the only one of its type in the world and is classified as a Bowstring Suspension Bridge. Until 1962, when it was reconstructed along more conventional lines, there was one other similar structure in existence, being the Chepstow Bridge, which carried the main railway line between Gloucester and South Wales. This was also designed and built by Brunel and opened to traffic in July 1852, in many ways this was the prototype for the Royal Albert Bridge.

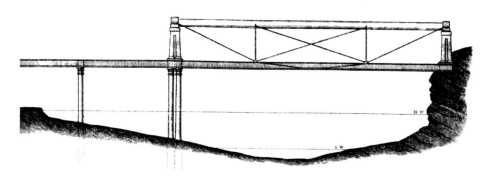

Drawing of Brunel's bridge across the River Wye at Chepstow.   H. Adlard.

Let us have a look at the design and construction of the trusses in greater detail. Basically, the passage of a train through the truss exerts a downward pull on the suspension chains which, in turn exert an inward pull on the chain anchorages in the tops of the piers. This is counteracted by the presence of the arched tubes between the piers.

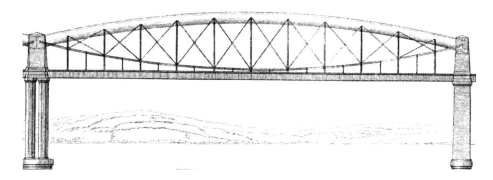

A side elevation of the Eastern Truss of the Royal Albert Bridge.   H. Adlard.

Due to the lack of published data, Brunel was forced to carry out experiments with different structural forms. He found that the strongest form of arch would be composed of tubes of oval cross section, giving the additional benefit of allowing the hangers for the deck to be hung vertically, thus reducing stress. Also, tubes of oval cross section would offer less resistance to the wind than the equivalent structure using tubes of circular cross section. At the time of construction, there was no available accurate data on wind forces, but one engineer, Smeaton, had presented a paper to the Royal Society in 1759 stating that wind forces varied from 6lb. per square foot for a high wind to 12lb. per square foot for a storm or tempest.

Each tube is 16 feet 9 inches wide and 12 feet 3 inches high and is constructed of wrought iron plates 10 feet long, 2 feet wide and between ½ - ¾ of an inch in thickness. They are stiffened internally by means of wrought iron plates and tie bars. At the ends of the tubes where the chains pass into their anchorages on the tops of the three piers additional plating is provided as the two opposite chains are the same distance apart from each other as the width of the tube, namely, 16 feet 9 inches.

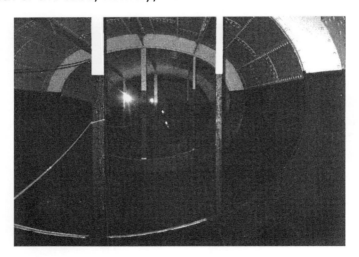

The interior of one of the tubes showing the internal stiffening plates and iron tie bars.
T.N. Bowden collection.

In addition, to prevent problems arising from the effects of condensation, two-inch diameter ventilation holes have been drilled throughout the length and breadth of the tubes.

20

The tubes are stiffened internally by means of six longitudinal stiffeners each made of plates, which are 12 inches deep. There are three on the top of the tubes and three on the bottom. In addition, there are 2-inch diameter iron tie rods, which run from the top to the bottom of the tubes, as well as flanged plates which run vertically from the top stiffeners to those on the bottom of the tube. The spacing of the tie rods is reduced towards the ends of the tubes. Annular stiffening plates, which are 15 inches deep are provided approximately every 20 feet. Additional stiffening has been provided where the two tubes meet over the central pier.

Eastern truss of the Bridge showing it's construction. The vertical hangers run from the bottom of the tube to the side girders at the bottom of the truss. Also note the diagonal links from the suspension chains to the side girders, these were installed in 1969 to strengthen the truss.
T.N. Bowden.

The decking is supported by means of hangers, which drop down from the tubes and by the suspension chains which are anchored at the tops of the piers. The tubes were designed in such a way that the hangers would drop vertically from the tubes to the side girders which support the decking in order to reduce the stresses, which occur when they are under a load.

The suspension chain links were originally manufactured by Messrs. Carne & Vivian of the Copperhouse Foundry, Hayle, Cornwall in about 1843 for use on the Clifton Suspension Bridge. However, the construction of that bridge was halted due to lack of funds and the chain links were sold to the Cornwall Railway Company for use on the Royal Albert Bridge. Upon purchase for the sum of £19 10/- (£19.50) per ton it was found that there were insufficient links for use on the Royal Albert Bridge, so some 1,650 additional links were supplied by Messrs Howard Ravenhill & Co. of the King and Queen Ironworks, Rotherhithe, London between 1856/7. These links differ slightly in their means of manufacture in as much as the links from the Copperhouse Foundry had welded lugs and those from Messrs. Howard Ravenhill & Co. were a one-piece forging. Up until 1845 it was only possible to weld the lugs onto the links. However, in 1845 Mr.Howard (of Howard, Ravenhill & Co.) took out a Patent for the forging of integral lugs onto chain links. The process involved two sets of rollers working at right angles to each other. This process resulted in stronger links and was altogether a more satisfactory arrangement. Brunel arranged for all of the links used in the Bridge to be proof tested on completion and this was done in the presence of one of his Assistants or Representatives.

The suspension chains consisted of two tiers; the centre of the chain's upper tier is 16 inches above that of the lower one. The chains alternate between 14 and 15 links. Each link is 20 feet long with a tolerance of only 0.02 inches and is 7 inches deep and either $^{15}/_{16}$th. inch or 1 inch thick. The links are held in place by means of 4-inch diameter pins, which pass through the lugs at the end of the links. Where the chains pass into the ends of the tubes, they are anchored by means of bolts 7 inches in diameter with a grip length of 24 inches. I should explain that where there are 15 links the individual links are $^{15}/_{16}$th inch thick and where there are 14 links each link is 1 inch thick.

Access to the tubes is provided by means of ladders and catwalks inside the landward ends of the trusses. For inspection and maintenance purposes it is possible to walk through the inside from one end of the tube to the other.

22

There are also two rails running the length of the top of the tubes. One of these rails is a toe rail which is only 3 inches high and was provided when the Bridge was built to enable ladders to be hung down the sides for repainting and repair purposes. The other rail was provided in the 1930's and is a 36 inch high handrail.

Relations between Brunel and the Contractor Charles John Mare started quite amicably however, Mare started to have problems having funds released by Brunel for the construction work that had already been completed. Relations between the two men became strained and eventually resulted in Mare going bankrupt in September 1855. The part completed works reverted to the Cornwall Railway Company and to their Engineer, Brunel, as per the original terms of the contract. This left Brunel in an unenviable position as all work on the Bridge ceased and he was forced to report to the Directors of the Cornwall Railway Company in March 1856 that this was owing to Mare's bankruptcy. Brunel went on to recommend that the Cornwall Railway Company should take over the contract under the direct supervision of his Chief Assistant, Brereton, who would be assisted by Gainsford, the Resident Engineer and George Whitting the Foreman. This recommendation was accepted and work on the Saltash or western truss restarted. The contract for constructing the Devon or eastern truss was let to Messrs. Hudson & Male. Mare's bankruptcy caused delays of nearly twelve months. There were further problems which resulted in delays to the completion of the eastern or Devon truss owing to adverse weather conditions.

After completion, the truss had to be tested before it could be floated out, so it was placed on temporary piers and a uniform load of 1,190 tons applied. When the midspan deflection was measured it was found to be only 5 inches, with the tests successfully completed, Brunel was able to plan the floating out and the lifting of the truss. Each truss including decking weighs around 1,060 tons and is 72 feet high from the top of the centre of the tube to the bottom of the decking.

Note the additional stiffening plates on the tube and how the plates are moved outwards where the ends of the tube and pier meet. *T.N. Bowden.*

Floating & Raising the Trusses

When the first truss had been completed, Brunel decided to modify both ends of the deck. He raised the two end panels of plate girders by just over 8 feet and moved them inwards towards the centre of the truss. This was achieved by disconnecting the vertical hangers under the chains from the side girders and holding the end panels in place by means of hawsers, which were passed through the tubes. This enabled the height of the truss above the water to be reduced, thus making it more manageable whilst being floated out. Docks were excavated beneath both ends of the truss, thus enabling the pontoons to be floated in and filled with water, before sinking onto specially constructed wooden cradles. The truss was then secured to the pontoons by means of a timber framework strengthened by iron tie rods. This completed, the pontoons were emptied at low tide and the valves closed, so that when the tide rose the pontoons became buoyant, thereby lifting the truss clear of the ground. The truss was then warped out onto the river before being positioned onto the temporary pier ends which had already been prepared to receive it.

Brunel decided to float out the western or Saltash truss on 1st. September 1857, but first he devised a way of giving instructions by means of coloured flags, 30 inches high and held in front of blackboards, which could be directed at the ship controlling the movement. The Master of the ship concerned had to give acknowledgement by repeating the signal.

The following code was drawn up:

Red - Heave in   White - Hold on   Blue – Pay out

Flags waved gently meant move gently, whilst flags waved violently meant move quickly.

Brunel had managed to borrow five naval vessels from Devonport Dockyard as well as a considerable amount of equipment needed for the floating out. This was in addition to the five hundred men from H.M.S. Ajax and Devonport. This supplemented his own staff who were assembled at their pre-determined posts on various vessels and pontoons. Local people also turned out as the day was declared a public holiday in the surrounding area. An estimated 40,000 crowd assembled on both sides of the river, many paying one shilling (5p) to watch the spectacle.

Shortly before midday, Brunel took up position on a specially constructed platform in the centre of the truss, where his assistant, Robert Brereton joined him and Captain Harrison. Brunel's friend, Captain Claxton, had been given the command of the five naval vessels involved. Robert Stephenson was expected, but was not able to attend due to a sudden illness.

Just before 1pm the message was relayed to Brunel that the truss had risen from the ground to the extent of three inches. He replied by giving the necessary signals to enable the huge truss to be warped out into the Tamar and then turned through ninety degrees, resulting in it being positioned over the ends of piers which had been built up to receive it. Whilst all this went on the crowds maintained an awe struck silence. By 3pm the truss had been secured without the slightest hitch and Brunel stepped down from the platform to the cheer of the crowd and the band of the Royal Marines playing 'See The Conquering Hero Comes' followed by "God Save The Queen". All looked well for a successful continuation of the project.

By July 1858 the western truss had reached its full height above the Tamar and the eastern or Devon truss was ready to be floated out. However, Brunel was not able to supervise this operation as his health had deteriorated and he was abroad recuperating. In his absence, Brereton supervised the movement, which was carried out without any problems.

The trusses were lifted into position by means of three hydraulic jacks placed under each end, although the middle jack on its own or the two outer ones, were capable of supporting the entire weight of the truss. Either using two 10 inch jacks together or a 20 inch jack on its own achieved each lift. A screw thread was cut into the rams of the jacks and a large threaded nut was kept tightly screwed up against the top of the jack as the ram emerged from it. This was to prevent the truss from falling should one of the jacks fail. As an additional safeguard against accidental movement, timber packing was placed between the bottom of the truss and the top of the masonry, which was being built up as the truss was lifted three feet at a time. Work progressed slowly in order to allow the masonry to set properly.

The centre pier comprises of four octagonal cast iron columns each 10 feet in diameter and using two inch thick plates with substantial internal stiffening. When completed, each column weighed 100 tons. The columns are positioned at 18 foot centres. Cross bracing was provided between each pair of columns but there was no longitudinal bracing. Each column was built up in sections as the truss was lifted.

Each truss took ten months to lift into its final position high above the Tamar. The raising of the trusses was made under the supervision of Brereton. No major problems were encountered during this long and difficult operation. During this time work progressed with the building of the land spans on both approaches. The landward end portals are supported on massive granite piers which measure 130 feet from their foundations to rail height. Each pier is 29 feet wide and is 17 feet thick. On the top of these piers are expansion bearings for the tubes, which are, made from 3½ inch thick wrought iron rollers. There are fixed bearings for the two tubes over the centre pier.

Before the Bridge could be opened for revenue earning traffic it had to be inspected by Colonel Yolland who was the Board of Trade Inspector. He inspected the Bridge over a three day period between 18th. – 20th. April 1859.

27

During his inspection of the Bridge a uniformly distributed load of 2¾ tons per foot run was applied over the entire length of the main spans and the midspan deflections were measured. The western span deflected by 7¾ inches and the eastern span by 7½ inches. In his report dated 25th. April 1859 he said that this was satisfactory and that the Bridge was one of the finest structures that he had inspected. His report for the line to Truro was not so glowing as he found that it required more ballast in places and the gradient boards had not been fitted. Remedial work was hastily undertaken and Colonel Yolland in his supplementary report dated 30th. April 1859 stated that the works had been carried out to a satisfactory standard and the line could now be opened for traffic.

One interesting point about the Board of Trade tests was that Colonel Yolland conducted a test of a train crossing the Bridge at 30 m.p.h. At this speed there was such severe oscillation in the main trusses that the spirit levels used for the tests fell off their mountings. In view of these results, Colonel Yolland imposed a speed limit on the bridge of 15 m.p.h. British Rail in the 1960's undertook further tests as they wished to raise the speed limit on the bridge and they encountered similar problems to Colonel Yolland in as much as at 30 m.p.h. there were undue oscillations within the main spans and that is why the speed limit still remains at 15 m.p.h.!

28

This picture shows one of the hydraulic presses used to lift the two trusses to their final position high above the river.   *T.N. Bowden collection.*

The western truss at its full height above the river around May 1858. The Eastern truss can be seen being built on the Devon shore.
W.J. Power collection.

A view from the Devon shore of the river after July 1858 showing the western truss at its full height above the river and the eastern truss secured on the pier ends. Note how the end panels of the deck have been lifted by 8 feet and secured to the tubes. This reduced the overall height of the truss above the water and made it more manageable whilst it was being floated out.
T.N. Bowden collection.

The granite top of the centre pier and the octagonal cast iron columns which support the inner ends of the trusses. The pier carries a sign reminding river users of the speed limit of 10 knots when passing under the Bridge *T.N. Bowden.*

THE BUILDING OF SALTASH BRIDGE

A cartoon from "Railway Ribaldry", a book published in 1935 to celebrate the Centenary of the Great Western Railway.
W. Heath Robinson.

The Opening Ceremony

It was in 1853 that H.R.H. Prince Albert, the Prince Consort, gave permission for his name to be given to the Bridge. He also agreed to perform the official opening ceremony on 2nd. May 1859. However, the first train ran from Plymouth to Truro on 11th. April, some three weeks previously. This train comprised of three coaches hauled by a South Devon Railway locomotive. It is believed that it was conveying Directors and Officials of both the Cornwall and South Devon Railway Companies. This train ran before the official inspection of the Bridge by Colonel Yolland.

On the day of the opening, special platforms were built at the Devon end of the Bridge for the Royal Party and at Saltash for the invited guests. At 12.30pm the Royal Train arrived from Windsor. Prince Albert did not leave the comfort of his carriage for the open platform. Instead he gave the Address from the steps of the Royal carriage! After the reply given by the Mayor of Saltash, Prince Albert returned inside for the crossing. A Battery positioned on the hill overlooking the Devon side of the river fired a salute when his train was halfway across the Bridge. The Royal Train continued slowly through Saltash station onto Coombe-by-Saltash viaduct where it was brought to a stand to allow a more panoramic view of the Bridge. After a while it was propelled slowly back into Saltash station where the Royal Party disembarked. Prince Albert then walked across the Bridge examining it in every detail, and upon reaching the Devon end went underneath to inspect the works and the plans.

The Prince later travelled on the Admiralty steamer "Vivid" to Torpoint, from where he travelled by road to Tregantle to inspect the new fortifications being built. After a short stay he returned to the "Vivid", which took the Royal Party to Millbay Docks. Then they went to Millbay Station where they enjoyed a substantial banquet before departing on the Royal Train for Windsor. Meanwhile, the invited guests were able to enjoy a cold collation laid on for their benefit in Saltash Guildhall.

Brunel was not present at the official opening ceremony as he was still abroad trying to recuperate from a serious illness, he was represented by Brereton. When Brunel eventually returned to England he was so ill that the only way to view his masterpiece was from a couch placed on a flat wagon drawn slowly across the Bridge by one of Daniel Gooch's locomotives.

Brunel died at his home in London on 15th. September 1859 and was buried in the family tomb in Kensal Green Cemetary, West London on 20th. September 1859. Victorian England had lost one of its best loved Engineers.

The Royal Train at Saltash station at the opening of the Royal Albert Bridge on 2nd. May 1859   *Illustrated London News*.

BRISTOL & EXETER RAILWAY.

VISIT

OF HIS ROYAL HIGHNESS

THE PRINCE CONSORT,

TO THE

OPENING

OF THE

ROYAL ALBERT BRIDGE,

AT

SALTASH,

ON

MONDAY, 2nd May, 1859.

ROYAL TRAIN TIME BILL.

DOWN.				UP.		
	DEP. A.M.	ARR. A.M.			DEP. P.M.	ARR. P.M.
WINDSOR	6 0		SALTASH		—	
Bristol		8 35	Cornwall Junction			—
"	8 45		"		6 50	
Taunton...		9 35	Newton			—
"	9 58		"		—	
Exeter		10 25	Exeter			8 15
"	10 35		"	8 25	
Newton		11 5	Taunton...			9 12
"	11 10		"	9 15	
Cornwall Junction		12 0	Bristol			10 5
"	12 5		"	10 15	
SALTASH		12 15	WINDSOR		12 50

*The following arrangements will be necessary for the proper working of
this Train, which must be strictly attended to:—*

The 7.50 a.m. Down Passenger Train is to Shunt at Tiverton Junction.
The 8.0 a.m. Goods Train Down will not start from Bristol until after the
Royal Train.
The 8.0 p.m. Up Train is to Shunt at Tiverton Junction.
The 9.20 p.m. Short Train from Weston is to Shunt at Yatton.

Bristol, 29th April, 1859.

Handbill by the Bristol & Exeter Railway Co. for the timings of the
Royal Train conveying H.R.H. Prince Albert to the opening of the
Royal Albert Bridge on 2nd. May 1859.   *T.N. Bowden collection.*

The Bridge
1859 - 2009

Over the years there have been various modifications made to the Bridge as well as routine maintenance. Some of the more notable works undertaken on the Bridge are detailed below:

Network Rail's New Measurement Train crossing the Bridge on 16th. May 2006.
Courtesy of Network Rail.

1859:
The lettering "I.K. BRUNEL ENGINEER 1859" placed on the portals at the end of both main spans in order to make the Bridge a permanent memorial to the late Isambard Kingdom Brunel. The lettering was supplied by the Plymouth Ironfounders for £15. The famous Coalbrookdale Company quoted £25 for the same work!

1867:
First full repaint of the Bridge at a cost of £1,700. The colours chosen were brown for the main spans and a granite colour for the four cast iron columns supporting the spans in the middle of the river. It is believed that the bridge was painted end to end every five years.

1868:
The lettering "I.K. Brunel Engineer 1859" on the portals at the landward ends of the two main spans were repainted at a cost of £2.

1892:
Over the weekend of 17th / 20th May the track on the bridge was converted from Broad Gauge, 7 feet 0¼ inches to the Standard Gauge of 4 feet 8½ inches. This was achieved by laying a Standard Gauge rail inside the Broad Gauge rails across the Bridge beforehand. Over the weekend of the gauge conversion the Broad Gauge rail were disconnected at either end of the Bridge and the Standard Gauge rail connected. The Broad Gauge rail was removed a few weeks later.

1905:
In 1905, 401 cross girders were replaced on the decking of the Bridge. It is interesting to look up at the two main spans from underneath as you will see that some of the cross girders run diagonally between the side girders whilst others are set at right angles. The cross girders on the land spans are all at right angles to the side girders.

1908:
The two land spans at the Saltash end of the Bridge were replaced in order to allow the double track through Saltash station to be extended partially onto the Bridge. This in turn led to improved station working and was the result of the growth of suburban traffic into Plymouth. The landspan nearest the end of the Down platform at Saltash has a Builders Plate on which reads, "E.Finch & Co Limtd Engineers & Ironfounders 1908 Chepstow".

A view of the underside of the Western Truss showing the cross girders which were replaced in 1905.   *T.N. Bowden.*

The two widened landspans at the west end of the Bridge which were installed in 1908 to improve the working of Saltash station.   *T.N. Bowden.*

1919:
The track across the bridge was originally laid on longitudinal timbers with cross sleepers to keep the track in gauge. In 1919 the track was relaid with conventional cross sleepered track without the longitudinal timbers.

1921:
Walkways were erected on the landward ends of both portals. Their purpose was to enable access to the bearing area and the internal parts of the tubes to allow inspection and maintenance of the Bridge to take place. Unfortunately the walkways partially covered the inscriptions on the landward ends of the portals.

1928/9:
In 1927, it was felt that with the increasing loads passing over the Bridge that the landspans on both sides of the Bridge should be replaced with new steel ones. At the same time the opportunity was taken to replace the cross girders and the original timber decking. Due to access problems as cranes could not be used, Swindon Works designed and built a special erection wagon, which was 95 feet long and weighed about 18½ tons. Temporary Construction Yards were established at both ends of the Bridge for the storage of equipment and materials.

Work was undertaken using a complete occupation of the line on several consecutive Sundays between 0900 – 1400. Trains ran from Plymouth to St. Budeaux where Road Motors were laid on to take passengers to the ferry at Saltash Passage.

On average, there were 46 men on site and it took them around 3½ hours to replace one span. The only two landspans, which were not replaced under this programme were the two that were replaced at the Saltash end of the Bridge in 1908.

Erection wagon built by the G.W.R. for the replacement of the landspans.
T.N. Bowden collection.

1930's:

A horizontal light lattice girder was added on each side of the two main spans. The purpose of this girder was to prevent the vertical hangers from bending when they were under a load. At the same time, additional cross bracing made from channel section steel was added to supplement the original wrought iron cross bracing used on the trusses.

It was during this period that an additional handrail was provided on the top of the two tubes. This rail is thirty-six inches high and compares with the toe rail's height of only three inches.

The top of the tubes looking towards Saltash showing both the hand and toe rails. *T.N. Bowden collection.*

Second World War:
During the early part of the Second World War, it was decided to lay a temporary timber decking across the Bridge for emergency military use should the need arise. Access to the Bridge was via a short access road off Vicarage Lane on the Devon side of the Bridge and off the north end of the 'Up' platform at Saltash station. Owing to problems caused by the position of the decking, the signalling equipment was modified to ensure the safe operation of the points and the Automatic Train Control ramps were moved. Trials were held with various military vehicles in case movement of vehicles and equipment had to be made across the Bridge. In order to protect the Bridge, gun emplacements were installed at both ends. The emplacement at the Devon end presented the Signalmen of Royal Albert Bridge Box with a practice shell at the end of the War, and this was kept in the Signalbox as a memento. The temporary decking was removed shortly after the end of the War.

1950's:
One plan for the Bridge, which was considered in the 1950's, was for the reinstatement of the timber decking removed after the War. This would have enabled cars to cross in between trains during the summer months to reduce delays on the Torpoint and Saltash Chain Ferries. There would have been traffic signals and level crossings at both ends of the Bridge and traffic would cross in one direction only between trains. Fortunately, Looe Urban District Council threw out this plan in 1959 as being unworkable.

1952:
The Bridge was listed as a Grade 1 structure. This means as it is of special architectural and historic interest that alterations can only be made after listed building consent has been obtained.

1959:
The Bridge had been specially prepared for its Centenary and had been repainted in 1958. Also, the walkways on the portals of the two main trusses had been temporarily removed to allow an uninterrupted view of the lettering, "I.K. BRUNEL ENGINEER 1859". From May until September the Bridge was illuminated at dusk by 264 lamps which were switched on every evening by the Signalman at Royal Albert Bridge Box. There were special boat cruises from Millbay Docks to the Bridge to see the illuminations. The old G.W.R. tenders "Sir Richard Grenville" and "Sir John Hawkins" were used to convey passengers up the river for 2/6d. (12½ pence) for adults and 1/3d. (6 pence) for children. The time and duration of the sailings depended on the time of sunset and the tides. Plymouth Corporation Transport provided special late buses from the docks to principal parts of the City for a set fare of 1 shilling (5 pence) for adults and 6d. (2½ pence) for children.

The Official Celebrations for the Centenary of the opening of the Bridge took place on 2nd. May 1959 and were as follows:

10.05am:	Special train to Saltash departs from Plymouth station.
10.15am:	Special train arrives Saltash station.
10.45am:	Assembly and Reception to be held in Saltash Guildhall.
11.15am:	Service of Thanksgiving to be held in Saltash Parish Church.
11.55am:	'Up' platform Saltash station - Unveiling of Brunel Centenary Plaque by the Mayor of Saltash, Alderman W.T.H. Stantake.
12.45pm:	Civic Luncheon in Saltash Guildhall.
2.45pm:	Special train departs from Saltash for Plymouth.
3.05pm:	Special train arrives at Plymouth.
3.15pm:	City Art Gallery, Plymouth - Royal Albert Bridge Centenary Exhibition to be opened by Sir John Carew Pole.
7.15pm:	Grand Hotel Plymouth - Civic Dinner, Lord Mayor of Plymouth to welcome guests.
9.45pm:	Special train departs Plymouth station for Saltash.
9.55pm:	Special train arrives at Saltash.
10.05pm:	Royal Albert Bridge floodlights to be switched on from station forecourt at Saltash by Mr. F. Edworthy, a member of British Railways Western Region engineering staff.
11.40pm:	Special train departs from Saltash station for Plymouth.
11.50pm:	Special train arrives at Plymouth.

The Brunel Centenary Plaque has been moved twice, firstly to the Devon end of the Bridge and then in 2006 to a location underneath the Bridge at Saltash.

As part of the Centenary celebrations, the Bridge was floodlit every night between May and September 1959.
T.N. Bowden collection.

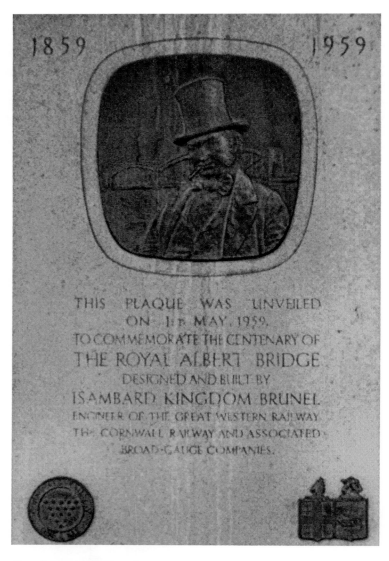

The plaque commemorating the Centenary of the opening of the bridge which is now located at Saltash. *Keith Hickman.*

1960:
In the middle of both mainspans the connections between the upper and lower vertical hangers were found to be suffering from corrosion where they passed between the links on the chains as they were inaccessible for maintenance. Temporary repairs had been made to keep the Bridge in service however, a more permanent solution was required. To solve the problem, new pin jointed hangers, were placed outside the chains and are connected directly to the decking side girders by means of special brackets.

1963:-
The Bridge was closed for several hours on 27th. October 1963 and again on 10th. November 1963, to enable the roller expansion joints at the free ends of the trusses to be renewed. A girder was affixed across the tracks at each end of the trusses in turn. This girder was raised up by means of two jacks who were nearly as old as the Bridge and had a lifting capacity of 45 tons each. They lifted the truss to the extent of one inch, the old roller bearings, a little flat in places, were removed and new sliding bearings installed. The old bearings, which were 12 inches long and 1½ inches in diameter, allowed for a movement of up to six inches, although, in fact, the movement was rarely more than two inches.

1966/7:
In the late 1960's with the impending modernisation of its freight rolling stock, British Rail conducted tests to see whether the Bridge would be able to carry the new generation of 50 foot long, 100 ton freight wagons which were about to be introduced. British Rail also wished to increase the speed limit of 15 m.p.h. across the Bridge, which had been, set by Colonel Yolland the Board of Trade Inspector in 1859. Tests carried out between November 1966 and April 1967 showed that at 30 m.p.h. lateral oscillation occurred which could in time damage the structure, therefore it was decided not to alter the speed limit. A perspex scale model, nine and a half feet long, was built at British Rail's Research Laboratory at Derby. Weights representing a train passing over could be hung on the model and strain gauges attached in order to determine the stress patterns in the structure.

As a result of these experiments, it was decided to insert further strengthening in the form of additional suspension links between the suspension chains and the decking, as this in turn would reduce the stresses in the vertical hangers when under load. The new links were made of steel $^5/_8$th. inch thick and 7 inches deep and placed diagonally between the chains and the decking from where, to a certain extent, they enhance the appearance when viewed from the side. The strengthening was carried out during 1969 at a cost of £80,000.

The additional diagonal chain links between the suspension chains and the side girders are clearly shown in this photograph of the eastern truss. Also visible is horizontal light lattice girder installed in the 1930's to reinforce the vertical hangers. *T.N. Bowden.*

Late 1960's:

Fears were being expressed about the lifespan of the Bridge. British Rail allayed these fears by holding a Press Conference where they stated that with proper maintenance the Bridge would remain in use for at least 25 years.

1978:

Once again fears were expressed in the late 1970's as to the Bridge's life expectancy. During November 1978, Mr. G.C. Skinner, the District Civil Engineer, stated that the structure would last well into the next century, since British Rail hoped to maintain the line into Cornwall for at least fifty years.

THIS PLAQUE WAS UNVEILED BY
HIS WORSHIP THE MAYOR OF SALTASH
COUNCILLOR ERIC LEWIS
ON 29TH APRIL 1984 TO COMMEMORATE
THE 125TH ANNIVERSARY OF THE OPENING OF
THE ROYAL ALBERT BRIDGE
BY H.R.H. THE PRINCE CONSORT
ON 2ND MAY 1859.

This plaque is on the 'down' platform at Saltash station. *T.N. Bowden.*

1987:

The timber supports on the insides of the main tubes were found to have rotted as a result of water ingress. Therefore it was decided to replace the timber with epoxy mortar on a rubber sheet as this would provide the same characteristics as the timber and would also prevent the ingress of water.

1997:

Railtrack Great Western Zone discovered on one of its regular inspections that the timber decking on the land spans required replacement. A £1 million contract was started in October 1997 which required the trackwork to be lifted and the timbers replaced. At the same time the opportunity was taken to to clean and examine the supporting steelwork before priming the girders and applying a new protective waterproof bitumastic coating. High tensile bolts were used to replace any rivets, which needed replacement. The new decking is made from Ekki a German Hardwood, which is long lasting and does not require treatment. The timber is laid longitudinally onto the newly protected cross girders and each piece of timber is 15 feet long by 12 inches wide and 6 inches thick. The new timbers are not bolted to the cross girders but rest upon them. The beams are butted up to each other and when they were all in place they were compressed together using jacks and packing pieces thus consolidating the new deck. The ballast boards were then replaced and the track and ballast replaced. The ballast used is only 4 inches deep and is made up of granite chippings with a nominal size of ½ inch. One downside of this is that the maintenance of the track has to be done by hand as the ballast is too shallow to be maintained by rail mounted machines.

2000:
Some of the timber bearings on the Western Truss needed replacement. The main contractor for this work was Dean and Dyball who in turn sub contracted the work to Elkspan Ltd. They came up with the solution of replacing the old timber bearings with new ones made from Oak and to Brunel's original design.

Elkspan Ltd. also retensioned the cross bracings, which had worked loose over the years. They designed a hydraulic tensioning system, which allowed the diagonals to be restrained and cut. The existing tension was measured and the cross bracings retensioned to the new specification.

The inspection of the inside of the portals took place in September 2000 when the Bridge had to be closed overnight to allow the inspection to take place.

In October 2000 the Bridge was examined in minute detail by two members of a three man team employed by Altus Access Ltd. They examined everything above the track including the vertical hangers and the portals over the track including the centre portal. Two members of the three man team used abseiling equipment! As part of the inspection they looked at the condition of the metal, corrosion, condition of the paintwork and checked every nut and bolt. The third member of the team conducted an internal inspection of the tubes.

Altus Access Ltd was working on behalf of Owen Williams Railways who in turn were contracted to do the work for Railtrack.

Mr. Martin, a spokesman for Owen Williams Railways said "It is such a high bridge that the most efficient way of getting access to all the awkward places is by using a professional team of rope access technicians." Mr. Martin also said, "The bridge is in an exposed situation where it is subjected to strong winds and sea spray."

The report on the condition of the bridge was passed to a Chartered Engineer who made recommendations to Railtrack who in turn decided which repairs should be made.

2002:

The cross sleepered track on the two main spans was replaced by track laid onto longitudinal timbers held in gauge by transoms as per Brunel's original design. This was done to try and make the decking more rigid when it was under load as movement in the decking was causing premature wear on the vertical hanger bearings. The ballast and planking underneath was replaced at the same time and additional restoration work was carried out on the Bridge. The work started at Easter 2002 and continued until the end of January 2003. The final phase of the work took place on the weekends of 11th. - 12th. January and 18th. - 19th. January when the Bridge was closed between Saturday morning and early Monday morning. There had been other short closures during 2002. In order to minimilise disruption to travellers, the works continued over Christmas Day and Boxing Day 2002. Over the Christmas period, there were 120 men working on the Bridge. The total costs of the restoration were £3 million with £1.2 million being spent on the Devon side of the structure and £1.8 million spent on the Cornish side. It is anticipated that the work will extend the lifespan of the Bridge by 25 years.

2006:
Network Rail moved the Centenary plaque from its position by the Tamar Bridge car park and relocated it to a position underneath the piers of the Royal Albert Bridge at Saltash. This was done as part of the Brunel 200 Birthday Celebrations. The ladders and walkways on the ends of the portals were removed and placed on the inner faces of the portals inside the main spans. This enabled the lettering "I.K. BRUNEL ENGINEER 1859" to be fully visible again. The unveiling took place on 16th. May 2006. Incidentally, this was the first time that the lettering has been fully visible since 1959.

2009 / 2010:
Network Rail who is responsible for the maintenance of the Bridge are intending to start an £8 million refurbishment programme which will see the two main spans stripped down to bare metal. Any necessary repairs will be made before painting with 3,750 litres of paint. The Bridge will require four coats of paint to give it proper protection from the elements, 1500 tins of paint will be used.

Devon Portal showing new walkways and ladders on the inside of the portal.   *T.N. Bowden.*

Signalling the Bridge

When the Cornwall Railway was first opened to traffic, the line was single throughout except for the first mile from Cornwall Junction to Devonport and a crossing loop provided at Saltash to allow two trains to pass. Constables were stationed at Devonport and Saltash and were responsible for working the disc and crossbar signals to protect the single line section. Permission for trains to proceed was given on the telegraph instruments. Later a signalbox containing a 20 lever frame was provided at Devonport in 1877 and a similar structure containing 18 levers opened at Saltash in 1882. The line was doubled between St. Budeaux and the eastern end of the Bridge on 23rd. February 1902 whilst at the same time, a new signalbox named 'Royal Albert Bridge' was brought into use to control the single to double line junction; the frame contained 13 levers. This signalbox had a short life for it closed on 28th. June 1908. At the same time, a new box containing a 15 lever frame was opened and a new Down Avoiding line was brought into use from St. Budeaux to Royal Albert Bridge. A new 25 lever frame was installed on 22nd. June 1952 when Royal Albert Box took over control of St. Budeaux West.

The line westward was doubled from Saltash to Wearde in 1906 and additional siding connections were installed at Saltash in 1909, along with a new 23 lever frame. In 1943 a new Down Goods Loop was brought into use between Saltash and Wearde and the box was extended with a new 31 lever frame.

Of the complete main line from London Paddington to Penzance, only half a mile has remained single tracked throughout its existence, that being the section over the Royal Albert Bridge. The considerable savings achieved by building the Bridge for a single line (£100,000 at 1859 prices) have often led to problems for the operating department, for although the single line section is only half a mile in length, it has a permanent speed restriction of 15 m.p.h. and each train could take two to three minutes to traverse the single line section. If the train was a heavy one, and had been stopped by signals at either end, it could take five minutes to clear the section due to the gradients leading onto the Bridge.

For many years, the single line was worked by the Electric Train Token system, which is an arrangement where only one token may be removed from holding machines by the joint action of the signalmen controlling the single line. The token is then given to the engineman as the authority to enter the single line section. The system is still in widespread use on British Rail.

The token working over the Bridge was abolished from 10th. July 1961 and replaced by a system of interlocking. In simple terms, the line appeared to be signalled as a double track main line and both signalboxes contained separate 'Up' and 'Down' normal double-line block instruments. An electrical current (known as a track circuit) was passed through the rails on the section of line between the two signalboxes and was interlocked with the block instruments. There was also an interlocking lever in each signalbox to control settings. This is a very rare method of working, particularly in the West Country, and it was very efficient over a short, busy section single line.

Changes came again on 2nd. July 1973 when the current mode of working was installed. The single line section from Saltash was extended to St. Budeaux and the Down Passenger Loop abolished. This was part of the extension of multiple aspect signalling controlled from the Panel Box at Plymouth North Road station. Saltash and Royal Albert Bridge Boxes were then closed. The latter has escaped demolition and today serves as the Network Rail Mobile Operations Manager's Office! Unfortunately, Saltash Box was demolished shortly after closure and a new British Rail 'bus stop' waiting shelter constructed on its site.

The Plymouth Panel Box now looks after the main line as far west as a fringe box at St. Germans and controls the priority of trains over the Royal Albert Bridge with a system known as 'track circuit block'. Entry to the Bridge is no longer controlled by the characteristic Great Western type signals and maybe a little of the romance has gone. No doubt, Brunel would have been impressed by a system, which allows trains to be passed over the Bridge from a control point over four miles away.

A Class 50 crossing the Bridge with an 'Up' passenger service bound for London Paddington. *T.N. Bowden.*

53

The Bridge Builders

This chapter deals specifically with the men and contractors employed during the design, building and testing of the Bridge.

A bust of Brunel at Saltash.
T.N. Bowden.

ISAMBARD KINGDOM BRUNEL:
Besides designing the Royal Albert Bridge, Brunel was appointed as the Engineer to the Great Western Railway (G.W.R.), between London and Bristol. As well as the G.W.R., Brunel was also involved in building various other railways both in this country and overseas.

Brunel was responsible for the design of three ocean going steamships, "Great Western", "Great Britain" (now preserved in Bristol) and the "Great Eastern". At the time of building, the "Great Eastern" or "Leviathan" as she was originally going to be called, was the largest ship in the world. She was 680 feet long with a beam of 83 feet. Her sheer size was not surpassed until early in the twentieth century and she could not pass through the new Suez Canal, as she was too wide! The "Great Eastern" was used for the laying of the first Transatlantic Telegraph cable, as she was the only vessel large enough to carry the cable and coal for the journey. For Bridgewater Docks Brunel designed a dragboat called "Bertha" which was used to remove silt from the Docks. "Bertha" is still in existence at a museum in Eyemouth, Scotland where she is the oldest working steamboat in the world. Brunel was also approached by the Admiralty and asked to undertake tests on their behalf to compare the efficiency of the screw propeller with vessels propelled by paddle wheels, "HMS Rattler" was put at his disposal for the tests.

During his short life, for he died at the age of 53, Brunel was also involved with his father's Thames Tunnel which became the world's first subterranean tunnel, the Gaz Engine which used carbolic acid and a portable military hospital for the Crimea. He was also on the committee for the Great Exhibition which took place in London and Vice President of the Instituition of Civil Engineers. Brunel was also sworn in as a Special Constable during the Bristol Riots in 1831.

Other projects included the Hungerford Suspension Bridge across the River Thames in London and the magnificent Clifton Suspension bridge across the Avon Gorge in Bristol. Unfortunately, he did not see the completed bridge, as it was not opened until 1864, five years after his death.

Like any other engineer he had his failures such as the Atmospheric Railway in South Devon and his designs for locomotives for the G.W.R. were absolute disasters. Yet, out of the latter adversity he employed a young locomotive engineer called Daniel Gooch who transformed steam locomotive design on the G.W.R.

Brunel was a family man, adoring his wife Mary and their three children, Isambard, Henry Marc and Florence Mary. Not that he was able to spend much time with them as he was often away from home and working over 20 hours a day. He designed a special carriage for his travels, which included special holders for his plans and notebooks as well as a special box for his cigars. He is reputed to have smoked up to 40 cigars a day, often his Assistants would find him asleep in a chair in the morning with the ash of a full cigar still on his clothes.

To summarise, Brunel was a great man and he had many good qualities, however he was driven by his own insecurity as he found it difficult to delegate and to be able to trust others. He was also a hard taskmaster with the Contractors who worked on various projects and was very slow at authorising payment for work done. On more than one occasion this caused a Contractor to declare themselves bankrupt.

There is a window in Westminster Abbey, London commemorating Brunel who was buried in the family tomb in Kensal Green Cemetary, West London on 20th. September 1859. In his memory his family and friends commissioned the sculptor Carlo Marochetti to make a statue which now stands in a prominent position on the Embankment in London.

Brunel's Chief Assistant, Robert Pearson Brereton.
T.N. Bowden collection.

ROBERT PEARSON BRERETON:
Brereton joined Brunel in 1836 as an Assistant at the age of 17. He made such an impression on Brunel that he was appointed to the position of his Chief Assistant in 1844 at the age of 25. On Brunel's death Brereton became his successor.

Most pictures of Brereton depict him with an eye patch, it is believed that he lost his left eye in an explosion in 1845. Brereton died in 1894 and there is a brass plaque commemorating him in the church at Blakeney, Norfolk as he was from a Norfolk family.

WILLIAM GLENNIE:
He was the Resident Engineer in 1849 when Brunel was making trial borings in the middle of the river whilst looking for a suitable rock for the foundations of the centre pier. Glennie proudly told his boss that he had stood on the rock underneath the Tamar and whilst he was there he had smoked a cigar! Prior to working at Saltash, Glennie was the Resident Engineer at Box during the construction of the tunnel, which took nearly 4 years to build using 3,000 men.

CAPTAIN HARRISON:
Captain William Harrison was born in 1813 and was the most distinguished captain of the Cunard Companys steamers when he was appointed as the first Master of the "Great Eastern". Captain Harrison became a close friend of Brunel advising him on matters concerning the building and equipping of the Great Eastern. Captain Harrison was drowned when his boat capsized whilst he was going ashore at Southampton from the "Great Eastern" on 21st. January 1860. His body is buried in St. James's Cemetary, Liverpool.

CAPTAIN CHRISTOPHER CLAXTON RN (ret'd):
Brunel first met Captain Christopher Claxton RN (ret'd) in Bristol where Claxton was the Quay Warden of the Port. Claxton became a close friend of Brunel advising him on the design of both the "Great Western" and "Great Britain" steamships. Claxton was Brunel's advisor on all maritime issues. Brunel invited Claxton to assist him with the floating out of the trusses at both Chepstow and Saltash.

As an aside, Christopher Claxton was interested in politics and was very pro-slavery. During the Bristol Riots of 1831, a rioter gained access to Claxton's house and was thrown out through a window by Claxton's black manservant!

CHARLES JOHN MARE:

C.J. Mare was a Shipbuilder who was the main contractor for construction of Stephenson's Britannia Bridge across the Menai Straits in North Wales. Mare had his own yard at Blackwall, London and he supplied the ironwork used in the construction of the Royal Albert Bridge. At the time of his bankruptcy in September 1855 his yard employed over 3,000 men. There was speculation at the time that Mare's gambling had caused his bankruptcy for he was famous for breeding and racing horses! Mare's creditors came to his rescue as he had orders for six gunboats from the Admiralty and the contract for the construction of Westminster Bridge in London. In 1857 Mare was trading again this time as The Thames Ironworks and Shipbuilding Company Limited. One of Mare's most famous vessels was the iron cylinder used for towing Cleopatra's needle from Egypt to England. The works had their own football team for their employees and this was called Thames Ironworks F.C. They decided to go professional in 1900 therefore, the original club was wound up and a new club was formed a month later. The name of the new club? West Ham United F.C.!

HOWARD, RAVENHILL & CO:

This was the Ironfounders used to produce additional links for the suspension chains after the original links had been purchased from the Clifton Suspension Bridge Company. The Copperhouse Foundry in Hayle, Cornwall, had made the original links around 1843. However, there were insufficient links and Brunel decided to use the services of the King & Queen Ironworks of Rotherhithe, London as Mr. Howard had patented a system of forging the lugs onto the links in one piece.

It is interesting to note that the King & Queen Ironworks of Rotherhithe were well known for producing high quality ironwork from previously used metal. Contemporary accounts of the methods used mention that "a number of poor women were used within the works to pile up the scrap iron into balls". These balls are then piled into a furnace for smelting. This produced high quality iron, which was known in the trade as King and Queen iron.

58

Messrs. HUDSON AND MALE:
Messrs. Hudson and Male completed the Devon truss after the bankruptcy of Charles John Mare. Unfortunately, I have not been able obtain any information about them.

COLONEL WILLIAM YOLLAND R.E. F.R.S.:
William Yolland was born in Plympton St. Mary in 1810 and was the son of John Yolland who was the Property Manager to Lord Morley of Plymouth. William Yolland's father encouraged his interests in surveying and land management and sent him to a school specialising in Mathematics. In 1828 he was commissioned into the Royal Engineers. After service in Britain, Ireland and Canada he was posted to the Ordnance Survey in 1838, a position he held until 1854 when he resigned and joined the Board of Trade's Railway Inspectorate as one of its Railway Inspectors. The Railway Inspectorate normally recruited it's staff from the Royal Engineers. In 1877 he was appointed to the post of HM Chief Railway Inspector. Col. Yolland died on 7th September 1885 in Atherstone, Warwickshire.

An aerial shot of the Royal Albert Bridge taken around the late 1920's.
T.N. Bowden collection.

Tales of the Bridge

Quite naturally, an engineering masterpiece such as the Royal Albert Bridge attracts some form of unofficial history of its own which is always worth recording for posterity.

It was said that the Cornish did not want the Bridge to be built, as it was feared that the Devil would use it to cross into Cornwall from Devon.

One source claimed that when the Bridge was completed, Brunel walked across it, and in sheer desperation, having realised that there was enough room for a double track, threw himself over the side into the River Tamar below.

Another source states that Brunel jumped off the platform, which had been specially constructed, at his request for the floating out of the first truss, and committed suicide.

One book makes reference to the fact that an enemy bomb during World War II hit the Bridge. However, the bomb failed to explode, and it was neatly deflected into the Tamar by one of the tubes. However, there is no evidence to substantiate this story, as subsequent examination of the tubes did not reveal any evidence of damage caused by being struck by a bomb or any other aerial bombardment device.

A lady who lives just below the Bridge in Saltash told me about the time in the mid 1970's when she looked out of her window only to see a boat going across the Bridge! Not only was it going across the Bridge but also she could not see anything in front of it! It turned out that the boat was on the end of a Motorail train and that the rest of the train was already within the two main spans. The lady who related this story to me must have been Cornish as she said that the train was going from Cornwall to England!

H.M.S. Roberts moored south of the Bridge at Saltash.
T.N. Bowden collection.

Once, a 'Monitor' class Warship, H.M.S. Roberts, tried to pass underneath the Devon span of the bridge at high tide, but lost the top six feet of its radio mast for its trouble. The Bridge was closed whilst a full inspection for any damage was made.

A Driver based at Plymouth Laira Depot once told of the time when he was a Fireman on one of the Auto-trains which ran from Plymouth to Saltash. He was passed on the Bridge by a pedigree cow, which made good its escape while being unloaded from a horsebox at Saltash station. It was eventually recaptured at St. Budeaux, some 1¼ miles away!

Another Driver, based at Laira, tells of the time when he was a Fireman to a Driver who always wanted to know where the ferry was, and in which direction it was sailing, whenever he crossed the Bridge. When asked for an explanation of this unusual question he said, "If the Bridge goes, wait until you are three feet from the water, jump, and swim for the ferry". One wonders what his escape route would have been after the ferry had been replaced by the Tamar Bridge!

Viewing the Bridge

Access to the Bridge is possible both on foot and by car to the riverbanks on both sides of the River Tamar. It is possible to photograph the Bridge from various vantage points. It is also possible to walk across the Tamar Bridge and view the Royal Albert Bridge at relatively close quarters as both bridges run parallel to each other and are only separated by about 200 feet.

If travelling by public transport it is possible to either travel to Saltash by rail or by bus from Plymouth. If travelling by rail, good views of the Bridge can be obtained from the ends of the platforms at Saltash. The photograph below was taken from the end of the 'Down' platform at Saltash station. And, of course, travelling by train to Saltash from Plymouth you will have the opportunity to examine the Bridge at close quarters as you pass over it at a stately 15 m.p.h.!

View of the Bridge from the end of the 'Down' platform at Saltash station.
T.N. Bowden.

62

The Bridge from the waterside at Saltash. *T.N. Bowden.*

Royal Albert Bridge on the left and the Tamar Bridge on the right as viewed from the Tamar Bridge Car Park on the Devon side of the river.
T.N. Bowden.

A view of the Bridge from Saltash Passage on the Devon bank of the river.
T.N. Bowden.

Epilogue

Brunel's first two designs for crossing the River Tamar were for double track bridges made from timber. His drawings for these designs do not appear to have survived and we can only speculate as to their appearance. Under the terms of its Act, the Cornwall Railway Company was only allowed to cross the River Tamar by means of a bridge of not more than four spans and Brunel had to get the approval of the Lord High Admiral for the design and materials used for its construction. The Admiralty were not happy with Brunel's timber designs as they then stipulated that there had to be a minimum height above high water of 100 feet and that there should not be more than one pier in the fairway. Another stipulation was that any bridge across the river had to have flat soffits. We know that Brunel intended to use a trussed arch as Brereton refers to it in his Paper to the Instituition of Civil Engineers in 1862, (please refer to page 7 of this book). I personally believe that Brunel was going to develop and enlarge the design which he used for his viaduct at Landore in South Wales. If Brunel had adopted such a design, he would have had to overcome the problem of finding a suitable rock on the river bed on which to construct the foundations for the piers. A drawing of Landore Viaduct is shown below:

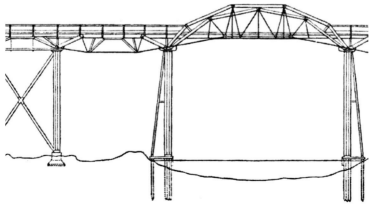

Brunel's Timber viaduct at Landore, Swansea.   *H. Adlard.*

We are also fortunate that Brunel chose wrought iron and not steel for the construction of the Bridge. Steel is not so resistant to corrosion as wrought iron and there is no doubt that a steel structure would certainly not have survived for 150 years.

The Bridge cost £225,000 to build in 1859, (£22.4 million at 2009 prices). At the time of completion this was a remarkably low figure compared to the costs of building similar structures elsewhere, especially considering the costs of building a deep water pier. I am sure that Brunel never envisaged celebrations being held in 2009 to commemorate the 150th anniversary of the opening of the Bridge!

The Bridge is now cared for by Network Rail who are responsible for all repairs and maintenance on the structure. Although it must be a difficult bridge to maintain, it is held in high esteem by the staff at Network Rail. There are no plans to close the Bridge in the foreseeable future, in fact, Network Rail are constantly looking at ways of prolonging it's life by improved methods of maintenance and by the use of modern materials.

Finally, in 1861, the Bridge carried 4,400 trains over a twelve month period, in 2006 this had increased to 21,000. The fact that the Bridge is capable of handling such an increase in traffic and of course heavier trains, is a tribute to its both its designer and to its custodians past and present who have maintained it.

The Bridge speaks for itself!

Materials used for the Construction of the Bridge
2,650 tons of Wrought Iron.
1,200 tons of Cast Iron.
459,000 cu. ft. of Masonry and Brickwork.
14,000 cu. ft. of Timber

1859 - "Hawk" a Broad Gauge saddle tank locomotive built in 1859 for the South Devon Railway (S.D.R.). Locomotives of this type were used to work trains on the Cornwall Railway.
T.N. Bowden collection.

2009 - A High Speed Train of the type which daily cross the Bridge on services to and from London Paddington. *T.N. Bowden.*

Bibliography

The following works were consulted during the preparation of this book. The author is indebted to the writers named below:

C.J. Allen –   "Great Western".
Ian Allan.

Keith Beck & John Copsey – "The Great Western in South Devon".
Wild Swan Publications.

Derrick Beckett –   "Brunel's Britain".
David & Charles.

Derrick Beckett –   "Stephenson's Britain".
David & Charles.

Stuart Berridge –   "Suspended by Chains".
Railway Magazine, November 1973.

Geoffrey Body –   "Clifton Suspension Bridge".
Moonraker Press.

Frank Booker –   "The Great Western Railway – a new history".
David & Charles.

Bradshaws -   "Bradshaws Railway Manual, Shareholders Guide and Directory 1869".
David & Charles reprints.

Isambard Brunel –   "The Life of Isambard Kingdom Brunel Engineer 1870".
David & Charles reprints.

Tim Bryan –	"Brunel The Great Engineer". Ian Allan.
Angus Buchanan –	"Brunel The Life and Times of Isambard Kingdom Brunel". Hambledon & London.
W.G. Chapman –	"The 10.30 Limited". Great Western Railway. Reprinted by Patrick Stephens.
W.G. Chapman –	"Track Topics". Great Western Railway. Reprinted by Patrick Stephens.
C.R. Clinker –	"New Light on the Gauge Conversion". Avon Anglia.
Ewan Corlett –	"The Iron Ship". Moonraker Press.
James Stevens Curl –	"Dictionary of Architecture". Oxford University Press.
Sally Dugan –	"Men of Iron". 4 Books.
Hamilton Ellis –	"Pictorial Encyclopaedia of Railways". Hamlyn.
C. Gaskell-Brown –	"Industrial Archaeology of Plymouth". Plymouth City Museum.
Peter Hay –	"Brunel Engineering Giant". Batsford.
W. Heath Robinson –	"Railway Ribaldry". Great Western Railway.

Luke Herbert –	"The Engineer's and Mechanic's Encyclopaedia". Thomas Kelly.
John Hunt –	"Refurbishing Cornwall's Gateway". "RAIL" – No. 327.
Edgar Jones –	"The Penguin Guide to the Railways of Britain". Penguin.
Andrew Kelly & Melanie Kelly -	"Brunel, In Love With The Impossible". Brunel Cultural Development Partnership.
Brian Lewis –	"Brunel's Timber Bridges and Viaducts". Ian Allan.
John Marshall –	"Guinness Book of Rail Facts & Feats". Guinness Superlatives.
O.S. Nock –	"The Rail Enthusiasts Encyclopaedia". Hutchinson.
O.S. Nock –	"The Great Western Railway in the 19th. Century". Ian Allan.
John Pudney –	"Brunel And His World". Thames and Hudson.
Sir Alfred Pugsley –	"The Works of Isambard Kingdom Brunel" (In particular the chapter entitled "Royal Albert Bridge" by Sir Hubert Shirley – Smith). Institution of Civil Engineers.
L.T.C. Rolt –	"Isambard Kingdom Brunel" Longmans Green.

David St. John Thomas –   "Westcountry Railway History".
　　　　　　　　　　　　　　David & Charles.

Richard Thames –   "Isambard Kingdom Brunel".
　　　　　　　　　　　Shire Publications.

A.C. Todd and Peter Law –   "Industrial Archaeology of Cornwall".
　　　　　　　　　　　　　　　　David & Charles.

David Walters –   "British Railway Bridges".
　　　　　　　　　　Ian Allan.

Archibald Williams –   "Brunel and After. The Romance of the Great Western Railway".
　　　　　　　　　　　　Great Western Railway. Reprinted by Patrick Stephens.

R.J. Woodfin –   "The Cornwall Railway to its Centenary in 1859".
　　　　　　　　　　Bradford Barton.

Adrian Vaughan –   "Isambard Kingdom Brunel Engineering Knight Errant".
　　　　　　　　　　　John Murray.

"Guide to the Great Western Museum, Swindon".

"New York Times" 8th September 1885.

"Western Morning News" various dates.

A side elevation of the Royal Albert Bridge.　　　　　　　　　　　　　　*H. Adlard.*

71

Acknowledgements

Firstly, I would like to offer special thanks to my wife Angela, whose patience and help has been invaluable in the preparation of this book.

I would also like to express my sincere thanks to the following individuals for so freely giving their time and assistance. Their help has been invaluable and through their efforts they have enriched this book.

Mavis Choong of Network Rail.
Vicky Stretch of Network Rail.
Peter Haigh of Network Rail.
Tom Holland of the Watermark, Ivybridge.
Sheila Butler.
Phillip Dredge.
Keith Hickman.
Simon Roseveare.
Mark Seaman.

I would also like to thank Devon Library and Information Services, particularly the ladies at Ivybridge Library for their assistance.

Special thanks are due to my old friend Bernard Mills for writing the chapter on "Signalling the Bridge".

Finally, I would like to thank anyone who I have inadvertently ommitted from the above list. Thank you!.